Program Overview Guide

Your handy reference to the supplements for your Holt science program

Because of the convenient organization of Chapter Resource Files, you can easily

- tailor instruction through a wide variety of leveled materials to meet individual student needs
- access a wide variety of engaging activities for students
- view all materials for a chapter in one tidy package
- emphasize process-skill development and practice with a range of well-defined lab types

CHAPTER RESOURCE FILE ORGANIZATION

The cover of each file provides a table of contents. Student copy masters are grouped together in the front of each file. Teacher materials are labeled "Teacher Resource Page" as shown above. Answer Keys are found at the end of each file.

TEACHING TRANSPARENCY LIST

To assist you in reviewing the program and in planning, titles for most of the transparencies related to each chapter appear on the front of each Chapter Resource File. Actual transparencies are provided as a separate package.

ABILITY-LEVEL LABELING

The materials in each file are categorized by ability level—basic, general, and advanced—to help you serve the needs of all students.

Basic Students at this level can manage core assignments only. Students who have special needs, struggling readers, and students who have certain learning disabilities fall into this category. See your Teacher Edition for additional help with instructional modifications for struggling students.

General Most students fall into this category. General-level students are still functioning at the concrete operational stage of intellectual development. They often need extra help to master abstract ideas and concepts.

Advanced Students at this level are able to handle honors or pre-AP coursework. These students are functioning at the formal operational stage of intellectual development. They can readily master the challenge of critical-thinking and synthesis activities.

ALSO ON THE ONE-STOP PLANNER®

All of the worksheets in the Chapter Resource Files are also located on the One-Stop Planner for your convenience.

Program Introduction

Program Resources Overview

Component		Description	Frequency
Skills Worksheets	Concept Review	These worksheets highlight and reinforce key concepts in each chapter.	one per chapter
	Critical Thinking	Students use cognitive skills to draw well-reasoned conclusions. Critical-thinking skills include distinguishing between fact and opinion; identifying bias; identifying fallacies in logic; judging the authenticity, worth, or accuracy of a position or claim; and synthesizing and evaluating information from various sources.	one per chapter
	Active Reading	Students learn to analyze text passages to build comprehension. The following skills are emphasized: recognizing cause and effect, sequencing, recognizing similarities and differences, and organizing information.	one per section
	Map Skills	Students use these worksheets to improve their map-reading and interpretation skills.	one per chapter
Assessments	Quizzes	Quizzes serve as quick comprehension checks that can be used to guide your reteaching options.	one per section
	Chapter Tests	Chapter Tests provide complete coverage of all chapter objectives in a simple and objective format.	two per chapter
Labs and Activities	Datasheets for In-Text Labs	Datasheets provide students with a copy of the textbook labs so that students do not need to take hardcover books into the lab. Charts, tables, and graphs are included to make data collection and analysis easier. Because space to write observations and conclusions is provided, the job of grading labs becomes much easier.	one per chapter
	Skills Practice Labs	Students practice standard laboratory techniques. These experiences are highly directed. Procedural steps guide students through the ideas needed to reinforce the concept of the lab.	variable by chapter
	Exploration Labs	Students explore a situation or phenomenon rather than explicitly testing a hypothesis. Although students often collect data in an exploration lab, the "end product" is a better understanding of a new concept or idea, as evidenced by a written analysis.	variable by chapter

Program Introduction

Basic use	General use	Advanced / enrichment use
Assign as homework or use as an in-class review of text concepts in preparation for the Chapter Test.	Assign as homework or use as an in-class study guide for pretest review.	Assign as homework to reinforce and review chapter concepts.
Optional: Because these worksheets are challenging, assign them as in-class peer-teaching activities.	Assign as homework or in-class group activities to teach critical-thinking processes.	Assign as homework to assess critical-thinking processes.
Assign as individual or peer-teaching group activities for struggling readers, or use to assess reading skill weaknesses.	Assign as homework for review and reinforcement of content and standardized test practice for reading skill development.	Optional: Assign as standardized test practice for reading skill development.
Assign as in-class group or peer-teaching activities.	Assign as homework or in-class group activities to reinforce map skills.	Assign as homework to reinforce map skills.
Determine gaps in understanding that require reteaching.	Determine whether students are keeping up with reading and assignments.	Determine whether students are keeping up with reading and assignments.
Assess student progress.	Assess student progress.	Assess student progress.
Datasheets help struggling students organize report information.	Datasheets provide a convenient option for lab reports on in-text labs.	Datasheets provide a convenient option for lab reports on in-text labs.
Using highly structured procedures and techniques, these labs provide concrete, manipulative experiences and models to help struggling students understand abstract or difficult concepts.	These labs give students opportunities to acquire and practice process skills and enable students to better understand text concepts.	These labs provide variety in daily lesson presentations and reinforce text content.
These labs provide concrete and manipulative experiences and models to help struggling students understand abstract or difficult concepts. These experiences enable you to use alternate modalities.	These labs provide manipulative experiences and concrete models to reinforce abstract concepts.	These labs provide variety in daily lesson presentations and reinforce text content.

Program Introduction

Component		Description	Frequency
Labs and Activities *continued*	Inquiry Labs	Students create and execute their own procedure to solve a problem. The problem is often presented as a real-life scenario in the form of a memo, letter, or description of steps. Students are provided with limited guidance and direction.	variable by chapter
	Modeling category	Students design and/or create a model to show their understanding of a concept, their ability to solve a problem, or their ability to answer an important question.	variable by chapter
	Field Activity category	Students perform experiments or collect data outside the classroom.	variable by chapter
	Long-Term Project category	Students explore a situation or phenomenon and collect data over a period of time (from as short as two weeks to as long as the school year).	variable by chapter
	Consumer category	Students apply concepts from the text when they examine and evaluate the use of natural resources and commercial products.	variable by chapter
	Observation category	Students make qualitative observations of living things or their environment.	variable by chapter
	Math/Graphing category	Students observe, record, calculate, and graph data. They draw conclusions by interpreting the data that they created.	variable by chapter
	Design Your Own category	Students design their own procedures to solve a problem or answer a question. These labs can involve additional research to build and test apparatus or a model with limited teacher supervision.	variable by chapter
	CBL™ Probeware category	Students use graphing calculators and probes or sensors to collect data. These experiments require some specialized equipment in addition to standard lab equipment. See p. 18 for more information.	variable by chapter
Teacher Resources	Lab Notes and Answers	The teacher's component for the Labs and Activities section is called Lab Notes and Answers. Components of Lab Notes and Answers are the following: • time required for the lab • process skills acquired for the lab • lab ratings, which provide information about the complexity of the lab in terms of teacher and student preparation, concept level, and cleanup required	included for each lab in the Chapter Resource File

Program Introduction

Basic use	General use	Advanced / enrichment use
Optional: Because these labs are complex, they should be used as group activities.	These labs help students develop process and analytical skills for standardized test preparation. Because these labs are complex, they should be used as group activities.	These labs provide the core experiences that challenge students at this level.
These labs provide simple manipulative opportunities to engage students.	These labs provide manipulative experiences and reinforce text content.	These labs provide variety in daily lesson presentations and reinforce text content.
Optional: To ensure safe experiences, assign these labs only as group activities.	Use these labs to provide solid experiences in data collection and analysis.	Use these labs to provide solid experiences in data collection and analysis.
Optional: Use these projects to develop writing, reference, and research skills.	Use these projects to develop writing, reference, and research skills.	Use these projects to develop writing, reference, and research skills.
These labs provide concrete and manipulative experiences and models to help struggling students understand abstract or difficult concepts. These experiences enable you to use alternate modalities.	These labs give students opportunities to acquire and practice process skills and enable students to better understand text concepts.	These labs provide solid experiences in data collection and analysis.
These simple experiences enable you to use alternate modalities to help struggling students better understand core concepts.	These labs give students opportunities to acquire and practice process skills and enable students to better understand text concepts.	Optional: These labs provide variety in daily lesson presentations and reinforce text content.
Optional: Because these labs are complex, they should be used as group activities.	These labs give students opportunities to acquire and practice process skills and enable students to better understand text concepts.	These labs provide solid experiences in data collection and analysis.
Optional: Because these labs are complex, they should be used as group activities.	These labs give students opportunities to acquire and practice process skills and enable students to better understand text concepts.	These labs provide the core experiences that challenge students at this level.
Optional: Because these labs are complex, they should be used as group activities.	These labs give students opportunities to acquire and practice process skills and enable students to better understand text concepts.	These labs provide solid experiences in data collection and analysis.
Leveling is provided only for student materials.	Leveling is provided only for student materials.	Leveling is provided only for student materials.

	Component	Description	Frequency
Teacher Resources *continued*	Lab Notes and Answers *continued*	• scientific method skills practiced in the lab • materials, safety, and disposal information • techniques to demonstrate • tips and tricks for making the lab a success • misconception alerts • answers to all questions	
	Answer Key for Skills Worksheets and Assessments	Detailed answers for skills worksheets and assessment items are provided in this section.	one section per Chapter Resource file
	Teaching Transparencies	Teaching Transparencies present text illustrations to highlight important concepts.	variable by chapter
	Bellringer Transparencies	Students engage in writing activities at the start of a section, which gives you time to complete administrative activities.	one per section
	Map Transparencies	Transparencies of important maps are available for large group instruction in map reading and interpretation.	variable by chapter

	Component	Description
One-Stop Planner	Chapter Resource Files	All worksheets, assessments, labs, and teacher resources for the entire program are found on the CD-ROM.
	Test Generator software	This test construction program enables you to provide customized tests by using the items in the Test Item Listing.
	Lab Materials QuickList software	This database lists all of the materials needed for the labs in your program. Using this software tool, you can print customized materials lists for the labs that you select.

Program Introduction

Basic use	General use	Advanced / enrichment use
Answer Keys are not leveled.	Answer Keys are not leveled.	Answer Keys are not leveled.
Teaching Transparencies are not leveled.	Teaching Transparencies are not leveled.	Teaching Transparencies are not leveled.
Bellringer Transparencies are not leveled. They should be used for in-class group activities.	Bellringer Transparencies are not leveled. They should be used for in-class group activities.	Bellringer Transparencies are not leveled. They should be used for in-class group activities.
Map Transparencies are not leveled. They should be used for in-class group activities.	Map Transparencies are not leveled. They should be used for in-class group activities.	Map Transparencies are not leveled. They should be used for in-class group activities.

Frequency
Chapter Resource Files for the entire program
all test item files for the complete program
all lab materials for the entire program

Safety with Microbes

WHAT YOU CAN'T SEE *CAN* HURT YOU

Pathogenic (disease-causing) microorganisms are not appropriate investigation tools in the high school laboratory and should never be used.

Consult with the school nurse to screen students whose immune system may be compromised by illness or who may be receiving immunosuppressive drug therapy. Such individuals are extraordinarily sensitive to potential infection from generally harmless microorganisms and should not participate in laboratory activities unless permitted to do so by a physician. Do not allow students who have any open cuts, abrasions, or open sores to work with microorganisms.

HOW TO USE ASEPTIC TECHNIQUE

- Demonstrate correct aseptic technique to students *prior* to conducting a lab activity. Never pipet liquid media by mouth. When possible, use sterile cotton applicator sticks instead of inoculating loops and Bunsen burner flames for culture inoculation. Remember to use appropriate precautions when disposing of cotton applicator sticks: they should be autoclaved or sterilized before disposal.

- Treat *all* microbes as pathogenic. Seal with tape all petri dishes containing bacterial cultures. Do not use blood agar plates, and never attempt to cultivate microbes from a human or animal source.

- Never dispose of microbe cultures without sterilizing them first. Autoclave or steam-sterilize at 120°C and 15 psi for 15 to 20 minutes all used cultures and any materials that have come in contact with them. If these devices are not available, flood or immerse these articles in full-strength household bleach for 30 minutes, and then discard. Use the autoclave or steam sterilizer yourself; do not allow students to use these devices.

- Wash all lab surfaces with a disinfectant solution before and after handling bacterial cultures.

HOW TO HANDLE BACTERIOLOGICAL SPILLS

- Never allow students to clean up bacteriological spills. Keep on hand a spill kit containing 500 mL of full-strength household bleach, biohazard bags (autoclavable), forceps, and paper towels.

- In the event of a bacterial spill, cover the area with a layer of paper towels. Wet the paper towels with bleach, and allow them to stand for 15 to 20 minutes. Wearing gloves and using forceps, place the residue in the biohazard bag. If broken glass is present, use a brush and dustpan to collect material, and place it in a suitably marked container.

   Program Introduction

Safety in the Laboratory

Systematic, careful lab work is an essential part of any science program. The equipment and apparatus students will use present various safety hazards. You must be aware of these hazards before students engage in any lab activity. The Lab Notes and Answers sections will guide you in properly directing the equipment use during the experiments. ***Holt Science: Laboratory Manager's Professional Reference*** contains detailed information regarding your responsibilities and liabilities as the lab manager.

Photocopy the following information for students. These safety rules always apply in the lab.

1. **Always wear a lab apron and safety goggles.** Wear these safety devices whenever you are in the lab, not just when you are working on an experiment.

2. **No contact lenses in the lab.** Contact lenses should not be worn during any investigations in which you are using chemicals (even if you are wearing goggles). In the event of an accident, chemicals can get behind contact lenses and cause serious damage before the lenses can be removed. If your doctor requires that you wear contact lenses instead of glasses, you should wear eye-cup safety goggles in the lab. Ask your doctor or your teacher how to use this very important and special eye protection.

3. **Personal apparel should be appropriate for laboratory work.** On lab days, avoid wearing long necklaces, dangling bracelets, bulky jewelry, and bulky or loose-fitting clothing. Long hair should be tied back. Loose, flopping, or dangling items may get caught in moving parts, accidentally contact electrical connections, or interfere with the investigation in some potentially hazardous manner. In addition, chemical fumes may react with some jewelry, such as pearls, and ruin them. Cotton clothing is preferable to wool, nylon, or polyesters. Wear shoes that will protect your feet from chemical spills and falling objects—no open-toed shoes or sandals and no shoes with woven leather straps.

4. **NEVER work alone in the laboratory.** Work in the lab only while under the supervision of your teacher. Do not leave equipment unattended while it is in operation.

5. **Only books and notebooks needed for the activity should be in the lab.** Only the lab notebook and perhaps the textbook should be used. Keep other books, backpacks, purses, and similar items in your desk, locker, or designated storage area.

6. **Read the entire activity before entering the lab.** Your teacher will review any applicable safety precautions before you begin the lab activity. If you are not sure of something, ask your teacher about it.

7. **Always heed safety symbols and cautions in the instructions for the experiments, in handouts, and on posters in the room, and always heed cautions given verbally by your teacher.** They are provided for your safety.

8. **Know the proper fire drill procedures and the locations of fire exits and emergency equipment.** Make sure you know the procedures to follow in case of a fire or other emergency.

| Safety in the Laboratory *continued*

9. If your clothing catches on fire, do not run; WALK to the safety shower, stand under the showerhead, and turn the water on. Call to your teacher while you do this.

10. Report all accidents to the teacher IMMEDIATELY, no matter how minor. In addition, if you get a headache or feel ill or dizzy, tell your teacher immediately.

11. Report all spills to your teacher immediately. Call your teacher, rather than cleaning a spill yourself. Your teacher will tell you if it is safe for you to clean up the spill. If it is not safe for you to clean up the spill, your teacher will know how the spill should be cleaned up safely.

12. Design Your Own and Inquiry Lab procedures must be approved by your teacher BEFORE you begin work.

13. DO NOT perform unauthorized experiments or use equipment or apparatus in a manner for which they were not intended. Use only materials and equipment listed in the activity equipment list or authorized by your teacher. Steps in a procedure should only be performed as described in the book or lab manual or as approved by your teacher.

14. Stay alert while in the lab, and proceed with caution. Be aware of others near you or your equipment when you are proceeding with the experiment. If you are not sure of how to proceed, ask your teacher for help.

15. Horseplay in the lab is very dangerous. Laboratory equipment and apparatus are not toys; never play in the lab or use lab time or equipment for anything other than their intended purpose.

16. Food, beverages, and chewing gum are NEVER permitted in the laboratory.

17. NEVER taste chemicals. Do not touch chemicals or allow them to contact areas of bare skin.

18. Use extreme CAUTION when working with hot plates or other heating devices. Keep your head, hands, hair, and clothing away from the flame or heating area, and turn the devices off when they are not in use. Remember that metal surfaces connected to the heated area will become hot by conduction. Gas burners should be lit only with a spark lighter. Make sure all heating devices and gas valves are turned off before leaving the laboratory. Never leave a hot plate or other heating device unattended when it is in use. Remember that many metal, ceramic, and glass items do not always look hot when they are heated. Allow all items to cool before storing them.

19. Exercise caution when working with electrical equipment. Do not use electrical equipment that has frayed or twisted wires. Be sure your hands are dry before you use electrical equipment. Do not let electrical cords dangle from work stations; dangling cords can cause tripping or electrical shocks.

20. Keep work areas and apparatus clean and neat. Always clean up any clutter made during the course of lab work, rearrange apparatus in an orderly manner, and report any damaged or missing items.

21. Always thoroughly wash your hands with soap and water at the conclusion of each investigation.

    Program Introduction

|Safety Symbols

The following safety symbols will appear in the instructions for experiments and activities to emphasize important notes of caution. Learn what they represent so that you can take the appropriate precautions. Remember that the safety symbols represent hazards that apply to a specific activity, but the numbered rules given on the previous pages always apply to all laboratory work.

- Never put broken glass or ceramics in a regular waste container. Use a dustpan, brush, and heavy gloves to carefully pick up broken pieces and dispose of them in a container specifically provided for this purpose.

- Dispose of chemicals as instructed by your teacher. Never pour hazardous chemicals into a regular waste container. Never pour radioactive materials down the drain.

- When using a burner or hot plate, always wear goggles and an apron to protect your eyes and clothing. Tie back long hair, secure loose clothing, and remove loose jewelry.

- Never leave a hot plate unattended while it is on.

- Wire coils may heat up rapidly during this experiment. If heating occurs, open the switch immediately and handle the equipment with a mitt.

- Allow all equipment to cool before storing it.

- If your clothing catches on fire, walk to the emergency lab shower and use the shower to put out the fire.

- Perform this experiment in a clear area. Attach masses securely. Falling, dropped, or swinging dropped objects can cause serious injury.

- Use a mitt to handle resistors, light sources, and other equipment that may be hot. Allow all pieces of equipment to cool before storing them.

- If a thermometer breaks, notify the teacher **immediately.**

- Do not heat glassware that is broken, chipped, or cracked. Use tongs or a mitt to handle heated glassware and other equipment because it does not always look hot when it is heated. Allow all pieces of equipment to cool before storing them.

- If a bulb breaks, notify your teacher immediately. Do not remove broken bulbs from sockets.

        Program Introduction

Safety Symbols *continued*

- Never close a circuit until the setup has been approved by your teacher. Never rewire or adjust any element of a closed circuit.
- Never work with electricity near water; be sure the floor and all work surfaces are dry.
- If the pointer on any kind of meter moves off scale, open the circuit immediately by opening the switch.
- Do not work with any batteries, electrical devices, or magnets other than those provided by your teacher.

- Do not eat or drink anything in the laboratory. Never taste chemicals or allow them to come into contact with your skin or eyes.
- If a chemical gets on your skin, on your clothing, or in your eyes, rinse it off immediately and alert your teacher.
- Do not allow radioactive materials to come in contact with your skin, hair, clothing, or personal belongings. Although the materials used in this program are not hazardous when used properly, radioactive materials can cause serious illness when used improperly.

- Tie back long hair, secure loose clothing, and remove loose jewelry to prevent them from getting caught in moving or rotating parts or from coming in contact with hazardous chemicals.

- Wear eye protection, and perform experiments in a clear area. Swinging objects can cause serious injury.
- Avoid looking directly at a light source. Looking directly at a light source may cause permanent eye damage.
- Use knives and other sharp instruments with extreme care.

- Always obtain permission before bringing any animal to school.
- Handle animals carefully and respectfully.
- Wear disposable polyethylene gloves when handling any wild plant.
- Do not eat any part of a plant or plant seed used in the lab.
- Wash hands thoroughly after handling plants or animals.

   Program Introduction

| Laboratory Techniques

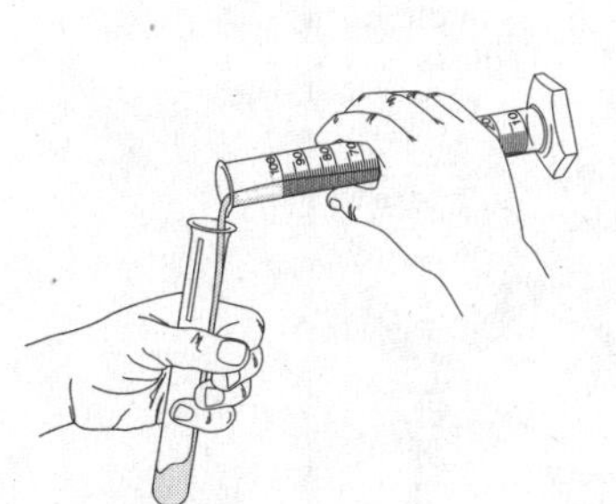

FIGURE A

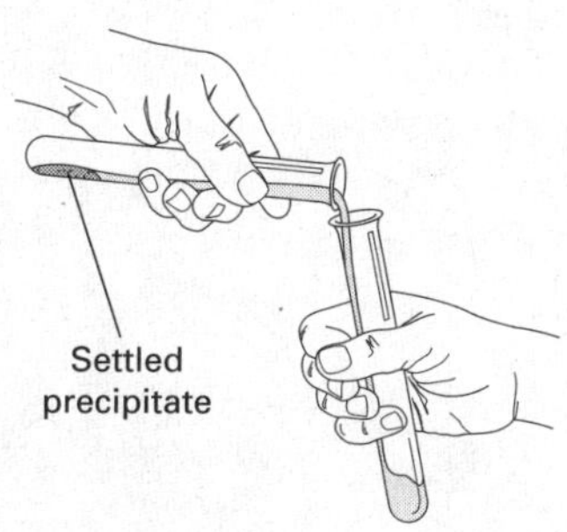

FIGURE B

FIGURE C

HOW TO DECANT AND TRANSFER LIQUIDS

1. The safest way to transfer a liquid from a graduated cylinder to a test tube is shown in Figure A. The liquid is transferred at arm's length, with the elbows slightly bent. This position enables you to see what you are doing while maintaining steady control of the equipment.

2. Sometimes, liquids contain particles of insoluble solids that sink to the bottom of a test tube or beaker. Use one of the methods shown above to separate a supernatant (the clear fluid) from insoluble solids.

  a. Figure B shows the proper method of decanting a supernatant liquid from a test tube.

  b. Figure C shows the proper method of decanting a supernatant liquid from a beaker by using a stirring rod. The rod should touch the wall of the receiving container. Hold the stirring rod against the lip of the beaker containing the supernatant. As you pour, the liquid will run down the rod and fall into the beaker resting below. When you use this method, the liquid will not run down the side of the beaker from which you are pouring.

HOW TO HEAT SUBSTANCES AND EVAPORATE SOLUTIONS

1. Use care in selecting glassware for high-temperature heating. The glassware should be heat resistant.

2. When heating glassware by using a gas flame, use a ceramic-centered wire gauze to protect glassware from direct contact with the flame. Wire gauzes can withstand extremely high temperatures and will help prevent glassware from breaking. Figure D shows the proper setup for evaporating a solution over a water bath.

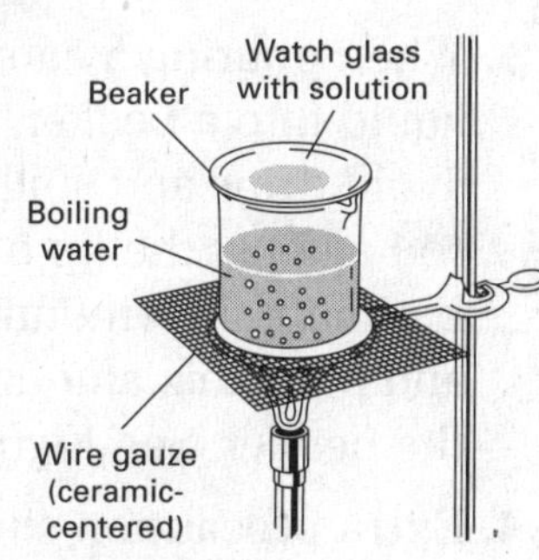

FIGURE D

3. In some experiments, you are required to heat a substance to high temperatures in a porcelain crucible. Figure E shows the proper apparatus setup used to accomplish this task.

4. Figure F shows the proper setup for evaporating a solution in a porcelain evaporating dish with a watch glass cover that prevents spattering.

Laboratory Techniques *continued*

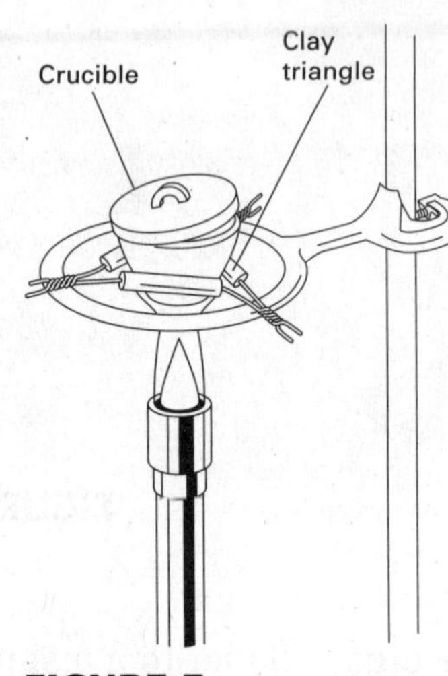

FIGURE E

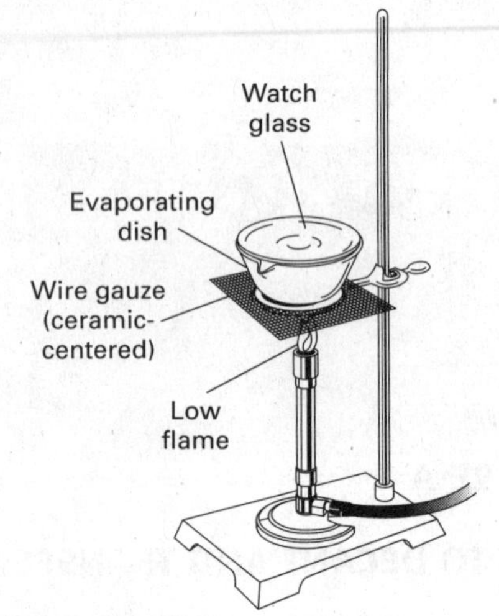

FIGURE F

5. Glassware, porcelain, and iron rings that have been heated may *look* cool after they are removed from a heat source, but these items can still burn your skin even after several minutes of cooling. Use tongs, test-tube holders, or heat-resistant mitts and pads whenever you handle these pieces of apparatus.

6. You can test the temperature of beakers, ring stands, wire gauzes, or other pieces of apparatus that have been heated by holding the back of your hand close to their surfaces before grasping them. You will be able to feel any energy as heat generated from the hot surfaces. DO NOT TOUCH THE APPARATUS. Allow plenty of time for the apparatus to cool before handling.

HOW TO POUR LIQUID FROM A REAGENT BOTTLE

1. Read the label at least three times before using the contents of a reagent bottle.

2. Never lay the stopper of a reagent bottle on the lab table.

3. When pouring a caustic or corrosive liquid into a beaker, use a stirring rod to avoid drips and spills. Hold the stirring rod against the lip of the reagent bottle. Estimate the amount of liquid you need, and pour this amount along the rod, into the beaker. See Figure G.

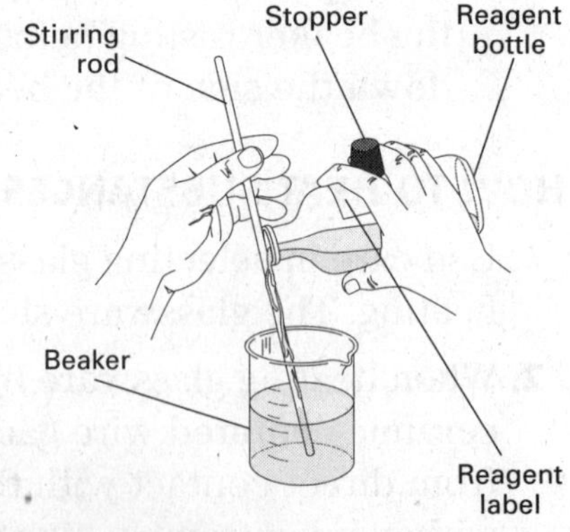

FIGURE G

4. Extra precaution should be taken when handling a bottle of acid. Remember the following important rules: Never add water to any concentrated acid, particularly sulfuric acid, because the mixture can splash and will generate a lot of energy as heat. To dilute any acid, add the acid to water in small quantities while stirring slowly. Remember the "triple A's"—*Always Add Acid* to water.

Program Introduction

5. Examine the outside of the reagent bottle for any liquid that has dripped down the bottle or spilled on the counter top. Your teacher will show you the proper procedures for cleaning up a chemical spill.

6. Never pour reagents back into stock bottles. At the end of the experiment, your teacher will tell you how to dispose of any excess chemicals.

HOW TO HEAT MATERIAL IN A TEST TUBE

1. Check to see that the test tube is heat resistant.

2. Always use a test tube holder or clamp when heating a test tube.

3. Never point a heated test tube at anyone, because the liquid may splash out of the test tube.

4. Never look down into the test tube while heating it.

5. Heat the test tube from the upper portions of the tube downward, and continuously move the test tube, as shown in Figure H. Do not heat any one spot on the test tube. Otherwise, a pressure buildup may cause the bottom of the tube to blow out.

HOW TO USE A MORTAR AND PESTLE

1. A mortar and pestle should be used for grinding only one substance at a time. See Figure I.

2. Never use a mortar and pestle for simultaneously mixing different substances.

3. Place the substance to be broken up into the mortar.

4. Pound the substance with the pestle, and grind to pulverize.

5. Remove the powdered substance with a porcelain spoon.

HOW TO DETECT ODORS SAFELY

1. Test for the odor of gases by wafting your hand over the test tube and cautiously sniffing the fumes as shown in Figure J.

2. Do not inhale any fumes directly.

3. Use a fume hood whenever poisonous or irritating fumes are present. DO NOT waft and sniff poisonous or irritating fumes.

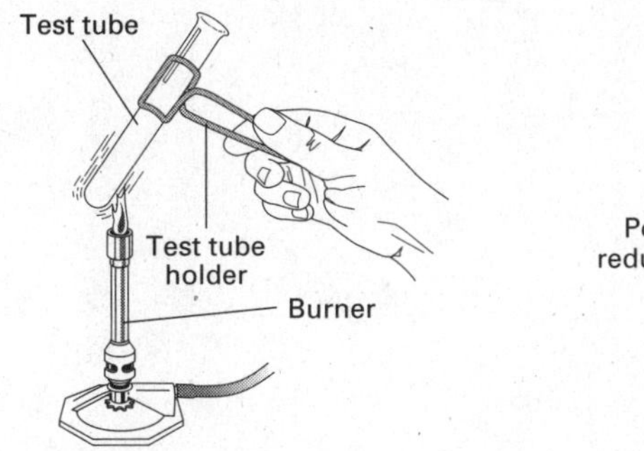

FIGURE H

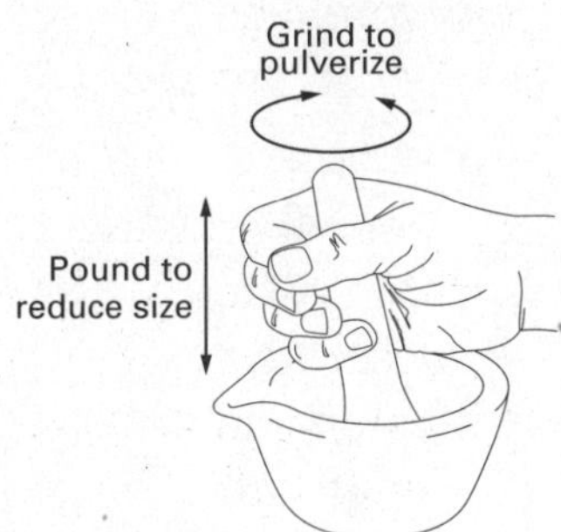

FIGURE I

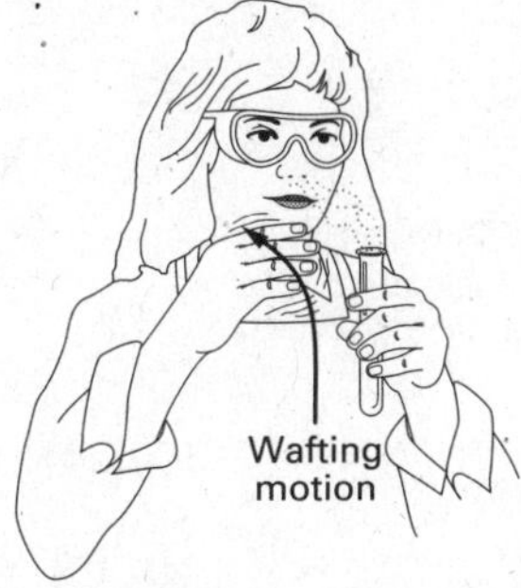

FIGURE J

   Program Introduction

Safety in the Field

Activities conducted outdoors require some advance planning to ensure a safe environment. The following general guidelines should be followed for fieldwork.

- **Know your mission.** Your teacher will tell you the goal of the field trip in advance. Be sure to have your permission slip approved before the trip, and check to be sure that you have all necessary supplies for the day's activity.

- **Find out about on-site hazards before setting out.** Determine whether poisonous plants or dangerous animals are likely to be present where you are going. Know how to identify these hazards. Find out about other hazards, such as steep or slippery terrain.

- **Wear protective clothing.** Dress in a manner that will keep you warm, comfortable, and dry. Decide in advance whether you will need sunglasses, a hat, gloves, boots, or rain gear to suit the terrain and local weather conditions.

- **Do not approach or touch wild animals.** If you see a threatening animal, call your teacher immediately. Avoid any living thing that may sting, bite, scratch, or otherwise cause injury.

- **Do not touch wild plants or pick wildflowers unless specifically instructed to do so by your teacher.** Many wild plants can be irritating or toxic. Never taste any wild plant.

- **Do not wander away from others.** Travel with a partner at all times. Stay within an area where you can be seen or heard in case you run into trouble.

- **Report all hazards or accidents to your teacher immediately.** Even if the incident seems unimportant, let your teacher know what happened.

- **Maintain the safety of the environment.** Do not remove anything from the field site without your teacher's permission. Stay on trails, when possible, to avoid trampling delicate vegetation. Never leave garbage behind at a field site. Leave natural areas as you found them.

Student Safety Contract

Read carefully the Student Safety Contract below. Then, fill in your name in the first blank, date the contract, and sign it.

Student Safety Contract

I will

- read the lab investigation before coming to class

- wear personal protective equipment as directed to protect my eyes, face, hands, and body while conducting class activities

- follow all instructions given by the teacher

- conduct myself in a responsible manner at all times in a laboratory situation

I, _________________________________, have read and agree to abide by the safety regulations as set forth above and any additional printed instructions provided by my teacher or the school district.

I agree to follow all other written and oral instructions given in class.

Date: _________________________________

Signature: _________________________________

Using Calculator-Based Labs

Download science lab software for FREE from Vernier Software & Technology or Holt, Rinehart and Winston to perform Calculator-Based Labs.

The procedures for the experiments labeled CBL™ Probeware (Calculator-Based Labs) are written for Calculator-Based Laboratory (CBL) Systems. There are a variety of systems currently available (such as CBL 1, CBL 2, and LabPro), and new products will continue to be developed (such as handheld devices that can be used with probes and sensors). Your equipment supplier is the best source of information regarding product updates or upgrades.

CBL Probeware experiments also require software programs by Vernier Software & Technology. In each experiment, the step-by-step instructions refer specifically to the menus that will appear on your students' calculator screens as they use these programs.

Texas Instruments offers a variety of graphing calculators that can be used with the CBL System. A list of these calculators is shown below. Although the instructions given in the experiments were written to work for all TI calculators, some minor adaptations may need to be made, depending on which calculators your students use.

- TI-73
- TI-82
- TI-83
- TI-83 Plus
- TI-86
- TI-89
- TI-92
- TI-92 Plus

Getting Started

SYSTEM REQUIREMENTS

You will need TI-GRAPH LINK™ software and a TI-GRAPH LINK computer-interface cable to get started using the PHYSCI programs with your TI graphing calculator and CBL. These products are available from CENCO, from Vernier Software, or from educational- and office-supply companies.

DOWNLOADING THE PROGRAMS FROM THE WEB TO YOUR COMPUTER

Use the following Web site to download the software programs you will need to complete Calculator-Based Labs.

---

internet connect

Go to: go.hrw.com
Keyword: HN4 VERNIER

At this site, you can download the lab software to your computer's hard drive. All download instructions are found on the Web site. After you download the software, open and read the text file. It contains more detailed information about the program by Vernier Software and Technology. You might want to print a copy of this file for reference throughout the school year.

You may also want to refer to the CBL Made Easy! Booklet that comes with your purchase of the CBL system. The most up-to-date version of the text file and this booklet can be found on the Vernier Web site at www.vernier.com.

    Program Introduction

Using Calculator-Based Labs *continued*

DOWNLOADING LAB SOFTWARE TO A CALCULATOR

- Each experiment will tell you which program you will need for that experiment. To download that group file from your computer to your calculator, follow the directions in the manual that comes with the TI-GRAPH LINK cable.
  Note: Because the program may be large, it is recommended that all other programs be removed before loading the program onto a calculator.

- When the transfer is complete, a set of programs will appear on the calculator's screen. The set will include the program title and a list of related programs that are used by the main program.

- To distribute the programs to your students' calculators, use the link-to-link cable to copy the programs to each student's calculator.

- Students should keep all of the programs on their calculators so that they can use them throughout the year. Because these programs are locked, students will not be able to change the programs on their own calculators.

USING THE SOFTWARE

- To use a program, simply follow the instructions given in the lab procedure. The program can calibrate sensors, collect data, and perform data analysis. You can use the CBL to its full potential without writing your own programs or using many different interfaces.

- The programs are designed to be very flexible. There are a variety of data-collection modes and sensors available. You even have the capability of using more than one sensor at a time so that several different measurements can be made simultaneously.

More Information

Texas Instruments

www.ti.com

Technical support: 1-800-TI-CARES

Programming assistance: (972) 917-8324

E-mail: ti-cares@ti.com

Address: Texas Instruments
P.O. Box 650311, MS 3962
Dallas, TX 75265

Vernier Software & Technology

www.vernier.com

E-mail: info@vernier.com

Address: Vernier Software &
Technology
13979 SW Millikan Way
Beaverton, OR 97005-2886

Program Introduction

Writing a Lab Report

There is no universally accepted format for a lab report. Your teacher will specify how he or she wishes to evaluate your work in the lab. However, the purpose of any lab report should be to record your findings and communicate what you have learned. With these goals in mind, here are some typical guidelines for writing a lab report.

Get organized. A good lab notebook is your key to success in the lab. Record everything in this notebook, and use it as a reference when you write the report.

Take notes during any pre-lab discussion. Good notes will make writing the report go much faster. You will usually be given some clues as to what to observe or how to do the calculations.

Always record the units when collecting data. It's easy to forget a few days later the units you saw for the data you collected.

MODEL FORMAT

A good general, scientifically based model has the following parts:

1. **Title** The title should clearly describe the nature of the experiment. In some cases, you may be able to use the title of the lab your teacher provides. However, be sure that the title provides clear information. Don't forget to include your name, date, and the names of any lab partners with your title section.

2. **Abstract or Summary** Though this section appears second in the report, it is often written last. This section summarizes the purpose of the experiment and your findings. It gives the reader a quick overview of what you've done and what you've learned.

3. **Introduction** This section should describe the problem or hypothesis you are investigating. The introduction should include the reason you are studying the problem and any useful outside information related to the problem or hypothesis.

4. **Materials and Methods** This section describes your procedure, the materials you used, how you gathered and analyzed your data, and the controls in your experiment. This section should be written in the past tense and passive voice.

For example:

Three 50 mL beakers were each filled with 25 mL of water.

Do not write:

I filled three beakers with water.

5. **Results** In this section, you describe what you found out through the experiment. Results include your observations, measurement data, graphs, and tables. Calculations and answers to all lab questions should be included.

Writing a Lab Report *continued*

DATA TABLES AND CHARTS

Choose a title for your data table, and then make a list of the types of data to be collected. This list will become the headings for your data columns. For example, if you collected data on plant growth over time, you could record your data in a table like the one below.

PLANT GROWTH OVER TIME

Time (days)	Height of plant (cm)
1	10
3	12
5	15
7	18
9	20

GRAPHS

Choose the scale for each axis of your graph. The scale should take up as much of the paper as possible so that the results can be clearly seen. Then, choose the interval for the scale (the number of days represented by each block in the x-axis scale, for example). Remember, once you choose the interval for the scale, you cannot change it. If you change the interval of your scale, your graph will not accurately represent your data.

Mark the points for each pair of numbers. When all points are marked, draw the best straight or curved line between them. Remember that you do not "connect the dots" when you draw a graph. Instead, you should draw a "best fit" curve—a line or smooth curve that intersects or comes as close as possible to your set of data points.

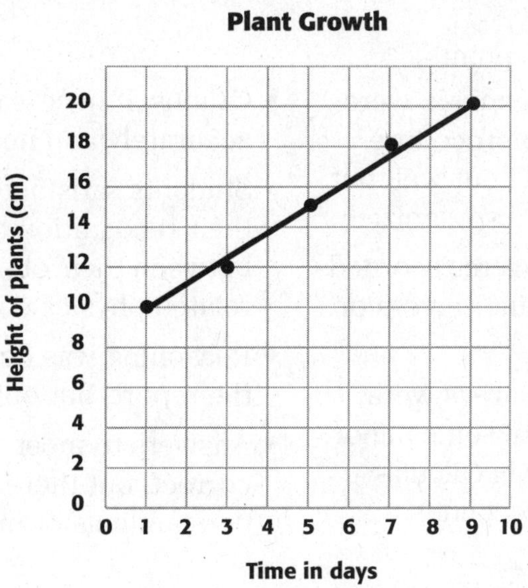

Scoring Rubric for Skills Practice Labs

The following rubric describes six levels of student performance in the laboratory. To use this 5-point scale, read the description of each level and decide which description most accurately reflects each report you grade.

EXPERIENCED LEVEL (5 points)

- Excellent technique was used throughout the lab procedure. Procedures were well-planned and well-executed.
- Data and observations were recorded accurately, descriptively, and completely, with no serious errors.
- Calculations and data analyses were performed clearly, concisely, and accurately, with correct units.
- Graphs, if necessary, were drawn accurately and neatly and were clearly labeled.
- Students recognized the connections between their observations and the related physics concepts; this understanding was expressed clearly and completely.
- Answers to questions were complete and were written correctly and accurately.

COMPETENT LEVEL (4 points)

- No errors in technique were observed during the lab procedure. Procedures were well-planned and were carried out in an organized fashion.
- Data and observations were recorded accurately, descriptively, and completely, with only minor errors.
- Calculations and data analyses were performed accurately, with correct units and properly worked-out calculations, but the work may have been slightly unclear or disorganized.
- Graphs, if necessary, were drawn accurately and neatly.
- Students effectively expressed their recognition of the connections between their observations and the related physics concepts.
- Answers to questions were written correctly and accurately but may have revealed minor misunderstandings.

INTERMEDIATE LEVEL (3 points)

- Only minor errors in technique were observed during the lab procedure. Procedures were carried out well but may have been slightly disorganized.
- Data and observations were recorded accurately, with only minor errors or omissions.
- Calculations and data analysis were performed accurately, but some minor errors were made either in calculations or in applying correct units.
- Graphs, if necessary, were drawn accurately and neatly.
- Students satisfactorily expressed their recognition of the connections between their observations and the related physics concepts.
- Reasoning was occasionally weak in the report, but only in a few places.
- Answers to most questions were correct, but there are some misunderstandings or minor errors.

Scoring Rubric for Skills Practice Labs *continued*

TRANSITIONAL LEVEL (2 points)

- Only a few errors in technique were observed during the lab procedure, but they may have been significant. Procedures may not have been well-planned, or they may have been carried out in a disorganized fashion.

- Data and observations were recorded adequately, with only minor errors or omissions.

- Calculations and data analysis were performed accurately, but minor errors were made both in calculations and in applying correct units.

- Graphs, if necessary, were drawn adequately.

- Students recognized connections between their observations and the related physics concepts, but this understanding was very weakly expressed.

- Reasoning was generally weak throughout the report.

- Some answers to questions were incorrect because of misunderstandings, minor errors, or poor data.

BEGINNING LEVEL (1 points)

- Several serious errors in technique were observed during the lab procedure. Procedures were not well-planned and were carried out in a disorganized fashion.

- Most data and observations were recorded adequately, but with several significant errors or omissions.

- Calculations and data analysis were performed inaccurately, but correct units were used most of the time.

- Graphs, if necessary, were drawn adequately.

- Students may not have recognized connections between their observations and the related physics concepts; no expression of understanding was evident in the report.

- Errors in logic were made in the report. The report may have been disorganized and unclear.

- Some answers to questions were incorrect or poorly written.

UNACCEPTABLE LEVEL (0 points)

- All work was unacceptable.

- No responses were relevant to lab.

- Major components of lab were missing.

Program Introduction

Scoring Rubric for "Design Your Own" Labs

For labs in which students are directed to develop their own procedures, it is essential for them to work in an organized and logical manner. Students must submit an initial plan for your approval before they begin work in the lab.

The following scoring rubric describes six levels of student performance in the laboratory to help you evaluate your students' lab work. Each level describes the organization and safety requirements for the initial plan, the methods and skills required in the lab, and the quality of analysis expected in the written lab report.

EXPERIENCED LEVEL (5 points)

- Plan showed careful and thorough planning with good reasoning and logic. Students expressed a clear understanding of the science concepts to be investigated through the plan.

- Plan was complete, appropriate, and safe.

- Proposed data tables were complete and clearly indicate all measurements that must be made to solve the problem.

- Excellent technique was used throughout the lab procedure.

- The final report followed the prescribed format. All apparatus was described in detail. All necessary diagrams, equations, and graphs were correctly labeled. The procedure and results were described clearly and in an organized fashion. Writing was clear, concise, and well-organized, with few grammatical or stylistic errors. The connection between the initial problem and the results of the lab was clearly expressed.

- Students were successful at solving the problem presented by the lab. Percentage error for quantitative answers was less than 15%.

COMPETENT LEVEL (4 points)

- Plan showed careful planning, although the reasoning and logic behind it may not have been clearly expressed. Plans reflected some understanding of the science concepts to be investigated through the lab.

- Plan was appropriate, safe, and nearly complete.

- Proposed data tables indicated all measurements that must be made to solve the problem, but there may have been some minor errors or omissions.

- No errors in technique were observed during the lab procedure.

- The final report followed the prescribed format. All apparatus was described in detail. All necessary diagrams, equations, and graphs were correctly labeled. The procedure and results were described clearly and in an organized fashion. Writing was clear, concise, and well-organized, with few grammatical or stylistic errors. The connection between the initial problem and the results of the lab was clearly expressed.

- Students were essentially successful at solving the problem presented by the lab. Percentage error for quantitative answers was less than 25%.

Program Introduction

Scoring Rubric for "Design Your Own" Labs *continued*

INTERMEDIATE LEVEL (3 points)

- Plan showed some logic, but the reasoning could have been more careful, more thorough, or more clearly expressed. Plans reflected understanding of the science concepts to be investigated through the lab, but not clearly.

- Plan was appropriate and safe, but there were some omissions.

- Proposed data tables indicated all measurements that must be made to solve the problem, but no provision was made for multiple trials.

- Only minor errors in technique were observed during the lab procedure.

- The final report followed the format. All necessary diagrams, equations, and graphs were included, but they may not have been complete. Apparatus was vaguely described. The procedure and results were described, but the writing was not clear or organized. There may have been serious grammatical or stylistic errors. Students understood the connection between the initial problem and the outcome of the lab.

- Students were somewhat successful at solving the problem presented by the lab. Percentage error for quantitative answers was less than 35%.

TRANSITIONAL LEVEL (2 points)

- Plan showed some logic, but not enough to completely solve the problem. Plan reflected understanding of the science concepts to be investigated through the lab, but not clearly.

- Plan was safe, but it included inappropriate procedures or omitted necessary steps. Plan may not have directly addressed the problem presented. Planned procedure will probably not work as written. The plan was poorly written or disorganized.

- Proposed data tables may not have included all measurements that must be made to solve the problem.

- Procedures may not have been well planned, or they may have been carried out in a disorganized fashion.

- The final report followed the format, but each section may not have been completely addressed. There were serious grammatical or stylistic errors. Students may have understood the connection between the initial problem and the outcome of the lab, but this understanding was not expressed in the report.

- Students' results only approximately addressed the problem presented by the lab. Percentage error for quantitative answers was less than 50%.

Scoring Rubric for "Design Your Own" Labs *continued*

BEGINNING LEVEL (1 point)

- Plan showed very little logic or understanding of what is required to solve the problem. Plan did not reflect understanding of the science concepts to be investigated through the lab.
- Plan may not have been completely safe. The plan was poorly written.
- Proposed data tables did not include all measurements that must be made to solve the problem.
- Several serious errors in technique were observed during the lab procedure. Students attempted to solve the problem by trial-and-error.

- The final report followed the format, but there may have been several omissions. There were serious grammatical or stylistic errors. Students did not understand the connection between the initial problem and the outcome of the lab.
- Students' results may not have adequately addressed the problem presented by the lab. Percentage error for quantitative answers was less than 65%.

UNACCEPTABLE LEVEL (0 points)

- All work was unacceptable.
- Major components of the plan were missing. The plan was completely illogical, unsafe, or completely irrelevant to the problem.
- Major components of lab were missing.

- Data and observations were incomplete and did not address the problem presented in the lab.
- The report did not address the problem presented in the lab. Percentage error for quantitative data was more than 80%.

Student Project Guide

Whether you're preparing for a long-term project that requires research or for one that does not, here are some tips that will help you successfully complete the project.

GENERAL TIPS

- **Choose a topic that you like.** You'll be working on this project for a while, so make sure you're exploring a subject you find interesting. Choose a topic that you want to know more about, or consider tackling a topic or question that puzzles you from your science class. If you're having trouble finding a topic, talk to your teacher. Your teacher will have strategies to help you find a good subject.

- **Start early!** If you start doing research early, you'll have more time to find information that you need. Starting early also gives you time to change topics if the project isn't working out as you planned. You'll also have more time to prepare an informative and attractive presentation or paper.

- **Set small goals for yourself within the project.** For example, if you have four weeks in which to do a project, you may want to assign yourself weekly goals. Small weekly goals are easier to accomplish than one big project, and will help you track your progress.

RESEARCH GUIDELINES

- **Don't believe everything you read.** Question the reliability of all of your sources. Government agencies, professional associations, museums, or known scientific journals are generally reliable sources. It is important to confirm information by using a variety of sources.

- **Keep track of your sources.** A bibliography is a list of sources used in writing a paper. Most teachers require bibliographies as part of the paper. Make sure you write down all the information your teacher requires for citing sources before you start taking notes. A detailed, well-organized bibliography makes your paper more credible. It shows how much work you've put into your research. It also provides a list of resources in case you, your teacher, or another classmate want to study the topic further.

- **Keep track of your notes.** It is important to know which source your information comes from. One strategy is to keep notes on color-coded note cards. Use colored index cards, or design your own system by using paper and markers. Always keep your notes together so you don't lose them. You may want to use a pocket folder or a three-ring binder to keep your notes in one place.

Student Project Guide *continued*

INTERNET RESEARCH—FINDING RELIABLE SOURCES

The Web provides a highly efficient way to do your research. With just a few clicks, you can access a vast amount of information. However, when you do research on the Web, avoid drawing conclusions before you've checked the information for reliability. Often, you can tell when a Web site contains bias or is opinion-based. Some sites look very convincing but contain information that is not supported by scientific evidence or experimentation. When you are uncertain of a source's reliability, consider the following criteria before you decide to use the information in your research.

CRITERIA FOR RELIABLE WEB SITES

- The authors make their case based on adequate evidence.
- The authors interpret the data cautiously.
- The authors acknowledge and deal with opposing views or arguments.
- The authors give a list of current sources that support their claims.

Some characteristics of unreliable Web sites require practice to identify. One trick is to look at the sources that are linked to the source you want to verify. These links can give you some idea of reliability. If your source is linked to a number of questionable sites, it's probably not a good source.

CHECKING SOURCE LINKS

- Paste the home page address in the Search box of your favorite search engine.
- Scan the results, and review the suitability of any questionable links.

CHARACTERISTICS OF UNRELIABLE WEB SITES

- The authors make extraordinary claims with little supporting evidence.
- The authors relate evidence based on personal experience instead of referring to controlled studies.
- The author appeals to emotion rather than logic.
- The authors misrepresent or ignore opposing views.
- The arguments are biased to support a political or economic agenda.
- The site is linked to sites that support a particular political or economic agenda.

    Program Introduction

Student Project Guide *continued*

SCIENCE FAIR PROJECTS

You can find a complete information packet for science fair projects Holt, Rinehart and Winston's Web site. Log onto **go.hrw.com,** and type in the keyword listed below.

Here are a few quick tips to consider if you are doing a science fair project.

- Brainstorm ideas for topics, and write down those ideas in your science notebook or journal.

- Keep all of your project notes in one place. This will make it easier for you to review the information later as you prepare your project display, make a presentation, write a paper, or answer questions.

- Follow the steps of the scientific method.

- Take careful notes during your preliminary research, and record your questions and hypotheses. Keep all of your experimental designs and data in your notebook or journal.

- Don't do anything dangerous or unethical. Be sure to obey the safety and experimental guidelines established by your teacher.

- Remember, even if your experimentation doesn't support your hypothesis, you will have learned something. Investigate why your hypothesis might have been wrong, and then explain your conclusions in your presentation.

INTERVIEW PROJECTS

- Schedule interviews well in advance. When you contact the person, explain who you are and why you want an interview. Before your interview, do some background research on the person you are interviewing.

- Go to the interview with a prepared list of questions, a notebook, and a pen to jot down your answers. If you want to make a video or audio recording of the interview, be sure to get permission from the person ahead of time.

- During the interview, remember that your prepared list of questions is just to help get you started. Be respectful, and don't be afraid to ask new or follow-up questions.

- After the interview, send a thank-you note to thank the person for his or her time.

     Program Introduction

Student Project Guide *continued*

PROJECT PRESENTATIONS

At this stage, you will turn all of your hard work into a project to share with other people. Whether you're writing a paper or an article, making an oral presentation, or presenting your results in a series of graphs and tables, make sure your presentation follows the guidelines set by your teacher. The following are some important questions to consider as you build your final presentation.

- **What medium should I use?** There are many issues to consider in deciding the best medium for presenting your research. For example, if you are choosing an audiovisual approach, do you have all of the equipment that you will need? Can you operate the equipment well enough to create a polished presentation? Or would it be more appropriate to choose a more traditional method of presenting scientific research, such as a paper, a poster, or an oral presentation? There are many more ways of sharing what you have learned other than those described here. Think carefully about which medium would be most appropriate.

- **What are the length requirements?** Do enough research to allow you to explain the topic thoroughly. If your paper should be four pages long, rewrite and revise it until it is four pages long. If your oral presentation should take 10 minutes, practice your oral presentation and make sure that it is 10 minutes long. If you find that you do not have enough information to fulfill the length requirements for your project, you may have to do more research. A good way to avoid this problem is to start with more research than you think you will need. You can cut out the extra or less important information later.

- **How do I present my work scientifically?** Remember that scientific writing is different from creative writing. Be sure to state the facts, and be clear about what information is factual and what is your opinion. Show how the evidence you gathered supports your conclusions. If you are using graphs or tables, make sure they are easy to read. Give details about how and where you got your information. Even if you are presenting your findings in a creative manner, you must document your sources and be able to explain the science behind your work.

Name _________________________________ Class _______________ Date ______________

| Writing a Research Paper

Process Overview

Use this page to record information about the process of developing your research paper.

Decisions Made Record the decisions that you make about each of the following parts of your research paper.

General subject of your paper _______________________________________

Specific subject of your paper _______________________________________

Purpose ____

Audience ____

Major sources of information ____

____

____

Thesis (one or two sentences describing your topic and what you plan to cover in

the paper) ____

____

____

Steps in the Process As you write your research paper, list the date when you complete each step in the writing process. If you return to a step at a later date, record that date as well.

Prewriting	__________	Revising	__________
Writing	__________	Proofreading	__________
Peer/self-evaluation	__________	Preparing final draft	__________
Evaluator's name	__________	Publishing	__________

   Program Introduction

| Writing a Research Paper *continued*

Prewriting

Use this page to make an early plan, or informal outline, for your research paper.

Exercise Complete the items listed below.

Title of paper ____

Introduction ____

____

____

Thesis ____

____

Body ____

____

____

____

____

____

Conclusion ____

____

____

Major sources ____

____

____

Writing a Research Paper *continued*

Research Questions

After you have limited your research topic, formulate some questions to guide your research. Make sure that all of the questions relate directly to the particular topic and focus that you have chosen.

Exercise List your research questions on the chart below. Use questions that begin with *who, what, where, when, why,* and *how.* As you find answers to your questions, note these answers on the chart. Also note the source of your answer.

Research Topic ____

Research question	Answer(s)	Source(s)
1.		
2.		
3.		
4.		
5.		

  Program Introduction

Developing the Body of the Paper

The body of a research paper is based on the materials that you gathered during
your research. Ideas presented in the body of the paper can be quoted directly,
paraphrased, or summarized. Practice writing a body paragraph for your research
paper by following the directions below.

Copy two related notes onto the "note cards" below.

Write a topic sentence for the paragraph based on these two notes. The topic sentence should be related to your thesis or should support your thesis.

____

____

____

Use the information from the notes above to write a paragraph for the body of
your paper. Use transitions such as *furthermore, as a result, on the one hand,*
and *in summary* to show the connections between ideas in the paragraph.
Continue on separate paper if you wish.

____

____

____

____

____

Program Introduction

Writing a Research Paper *continued*

Evaluating and Revising

Evaluating Based on each of the following questions, rate the research paper (your own or that of a classmate) on a scale of **1** to **4,** with **1** being the lowest rating and **4** being the highest. For each rating that is less than 4, explain how the paper might be revised to improve that rating.

1. Does a thesis appear early in the paper?   1   2   3   4

____

2. Is the paper developed with sufficient sources that   1   2   3   4
    are relevant, reliable, recent, and representative?

____

3. Is the paper clear, interesting, and suitable for   1   2   3   4
    its audience?

____

4. Is the tone of the paper appropriate?   1   2   3   4

____

5. Are ideas and information stated mainly in the   1   2   3   4
    writer's own words?

____

6. Does all the information relate directly to the topic   1   2   3   4
    and thesis?

____

7. Has proper credit been given for each source of   1   2   3   4
    information?

____

8. Does documentation follow the style recommended   1   2   3   4
    by your teacher?

____

Revising Record at least two changes that you have made on your rough draft in response to the above evaluation.

____

____

____

____

      Program Introduction

Writing a Research Paper *continued*

Proofreading

Proofreading 1 In the space below, record any problems that you find in the areas of grammar, usage, spelling, capitalization, punctuation, and manuscript form (including the citation form in the body of the paper and the list of works cited at the end of the paper).

Grammar and usage	Description of problem	Location (page and line)

Spelling	Misspelled words	Correct spellings
Capitalization	**Words with errors**	**Words with errors corrected**

Punctuation	Description of problem	Location (page and line)
Manuscript form		

Proofreading 2 Record two changes that you made in your paper based on the information in the above charts.

____

____

____

Program Introduction

Writing a Research Paper *continued*

Process Review

After you have finished your research paper, write a paragraph describing what the easiest or hardest part of completing the paper was and why. Then, list one lesson that the assignment taught you about prewriting, writing, revising, proofreading, or publishing.

____

____

____

____

____

____

____

____

____

____

____

____

____

____

____

____

____

____

Rubric for Writing Assignments

POSSIBLE POINTS	CRITERIA
90–100	The assignment is engaging, concise, and polished. An attention-grabbing headline and a clear topic sentence in the first paragraph introduce the reader to the subject. Factual details, visual elements, quotations, and/or the proper use of scientific terminology add clarity and interest to the assignment.
80–89	The assignment is well written. An attention-grabbing headline and a clear topic sentence in the first paragraph introduce the reader to the subject. Factual details, visual elements, quotations, and/or the proper use of scientific terminology add clarity and interest to the assignment. Sometimes, the writing is slightly repetitive or unclear, but the writer demonstrates a good understanding of the subject matter.
70–79	The assignment is fairly well written and clear, but several errors indicate that the author may not have a complete understanding of the subject. Factual details, visual elements, quotations, and/or the proper use of scientific terminology are used.
60–69	The assignment has several significant problems in style and content. The topic is never clearly stated, scientific terms are misused, and misspelled words are present. Inadequate or incorrect use of factual details, visual elements, or quotations seems to indicate that the author does not have a solid understanding of the subject.
10–59	Although the assignment has been attempted, the author has clearly not put forth much effort. The writing is unclear, unfocused, and vague. A topic sentence is not given or is not adequately supported with details. The author has not used scientific terminology correctly and has introduced false statements and errors in writing style that make reading this work very difficult.
0	No work was completed.

POSSIBLE POINTS	SCIENTIFIC THOUGHT (40 points possible)
36–40	complete understanding of topic; topic extensively researched; variety of primary and secondary sources used and cited; proper and effective use of scientific vocabulary and terminology
31–35	good understanding of topic; topic well researched; a variety of sources used and cited; good use of scientific vocabulary and terminology
26–30	acceptable understanding of topic; adequate research evident; sources cited; adequate use of scientific terms
21–25	poor understanding of topic; inadequate research; little use of scientific terms
0–20	lacks an understanding of topic; very little research, if any; incorrect use of scientific terms

Program Introduction

Rubric for Presentations

POSSIBLE POINTS	**ORAL PRESENTATION** (30 points possible)
27–30	clear, concise, engaging presentation that is well supported by use of multisensory aids; scientific content effectively communicated to peer group
23–26	well-organized, interesting, confident presentation that is supported by multisensory aids; scientific content communicated to peer group
19–22	presentation acceptable; only modestly effective in communicating science content to peer group
16–18	presentation lacks clarity and organization; ineffective in communicating science content to peer group
5–15	poor presentation; does not communicate science content to peer group

POSSIBLE POINTS	**EXHIBIT OR PRESENTATION FORMAT** (30 points possible)
27–30	layout and presentation format that is self-explanatory and successfully incorporates a multisensory approach; creative use of materials
23–26	layout and presentation format that is logical, concise, and can be followed easily; materials used are appropriate and effective
19–22	layout and presentation format that is acceptable; materials are used appropriately
16–18	layout and presentation format that could be improved; somewhat ineffective use of materials
5–15	layout that lacks organization and is difficult to understand; poor and ineffective use of materials

Program Introduction